Practice Exam for the Civil PE Examination

Breadth Exam (40 questions)

Indranil Goswami, Ph.D., P.E.

September 2016

First printing September 2016

Preface

In January 2015, the official (NCEES) syllabus for the PE-Civil examination underwent a significant realignment. There was a significant departure from the previous structure which placed approximately equal emphasis on the five areas of practice. In the breadth exam, the 40 questions were approximately equally distributed among Construction, Geotechnical, Structural, Transportation and Water Resources. In the current syllabus for the Breadth (AM) exam, Transportation has been significantly deemphasized while there seems to be more emphasis on Construction.

The new depth (PM) syllabi have also gone through reorganization as well as addition of specific subtopics under various categories.

This (breadth) practice exam was developed *after* the most recent syllabus update (June 2016) and is therefore consistent with the same.

This practice exam contains 40 breadth (AM) questions. It should be taken under as near exam conditions as possible, preferably at the point when you think your exam review is complete and you are ready to take a simulated test to assess the level of your preparation. You should even go so far as to ask someone else to detach the questions from the solutions, so that you don't have any temptation to peek.

All the best for the upcoming PE exam,

Indranil Goswami

September 2016

Table of Contents

BREADTH EXAM QUESTIONS AM001 - AM040   05-25

BREADTH EXAM ANSWER KEY   28

BREADTH EXAM SOLUTIONS AM001 -AM 040   27-38

BREADTH EXAM
FOR THE
CIVIL PE EXAM

The following set of 40 questions (numbered AM001 to AM040) is representative of a 4-hour breadth (AM) exam according to the syllabus and guidelines for the Principles & Practice (P&P) of Civil Engineering Examination (updated July 2016) administered by the National Council of Examiners for Engineering and Surveying (NCEES). The exam is weighted according to the official NCEES syllabus (2016) in the following subject areas – Construction, Geotechnical, Structural, Transportation and Water & Environmental. Copyright and other intellectual property laws protect these materials. Reproduction or retransmission of the materials, in whole or in part, in any manner, without the prior written consent of the copyright holder, is a violation of copyright law.

The time allocated for this set of questions is 4 hours.

AM001

For the plane truss shown below, the force (lbs) in member BE is most nearly:
- A. 27,000 (compression)
- B. 27,000 (tension)
- C. 41,000 (compression)
- D. 41,000 (tension)

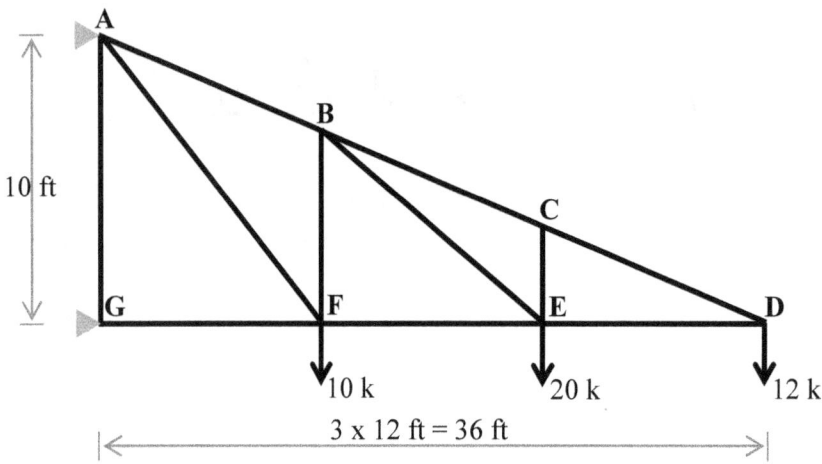

AM002

A rectangular beam is bent into an L-shape as shown below. A force P = 10 kips acts on the beam as shown. The maximum normal stress (lb/in^2) is most nearly
- A. 8,120
- B. 8,630
- C. 9,100
- D. 9,370

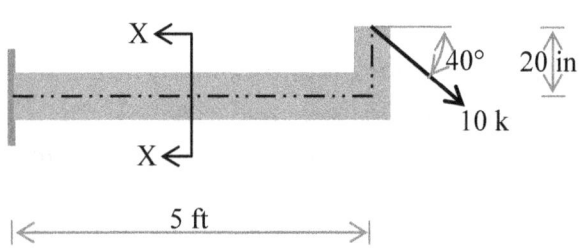 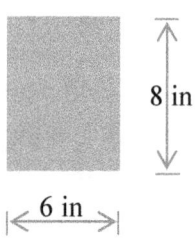

Section X-X

AM003

A beam is loaded as shown below. Which of the following represents an appropriate bending moment diagram for the beam?

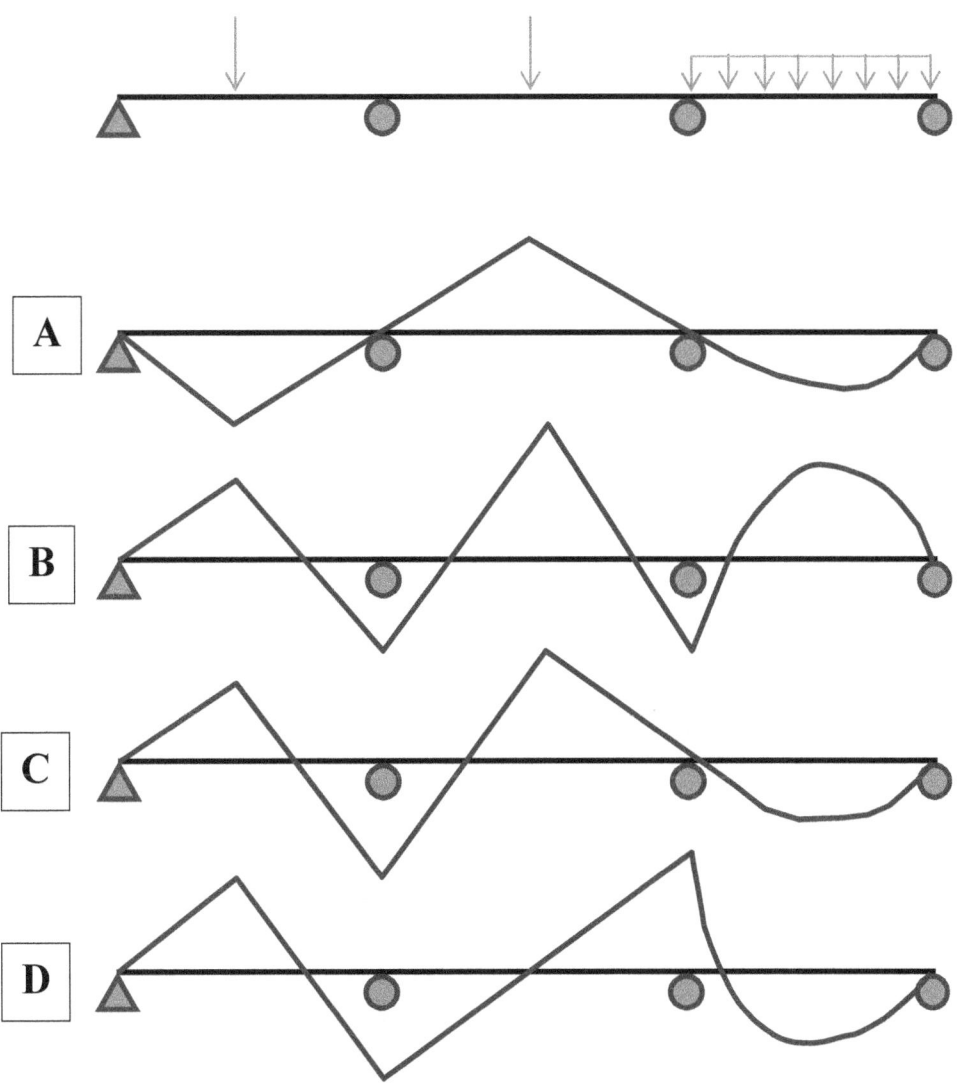

AM004

A cantilever beam is loaded at the free end as shown. Section properties of the doubly symmetric I-section are shown below the diagram. The maximum vertical deflection (inches) is most nearly:

    A. 1.23
    B. 1.54
    C. 1.90
    D. 2.32

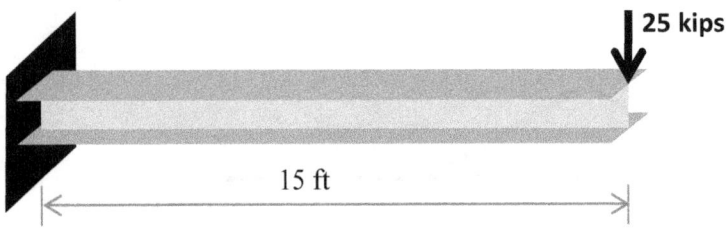

$A_g = 21.2$ in^2; $d = 11.7$ in; $I_x = 890$ in^4; $I_y = 235$ in^4; $S_x = 152$ in^3; $S_y = 70$ in^3

AM005

A rigid structural sub-assembly (weight = 12000 lb) is being lifted using two steel cables as shown below. Properties of the cables are shown. The vertical displacement (inches) of the center of gravity (CG) is most nearly:

    A. 0.12
    B. 0.15
    C. 0.18
    D. 0.22

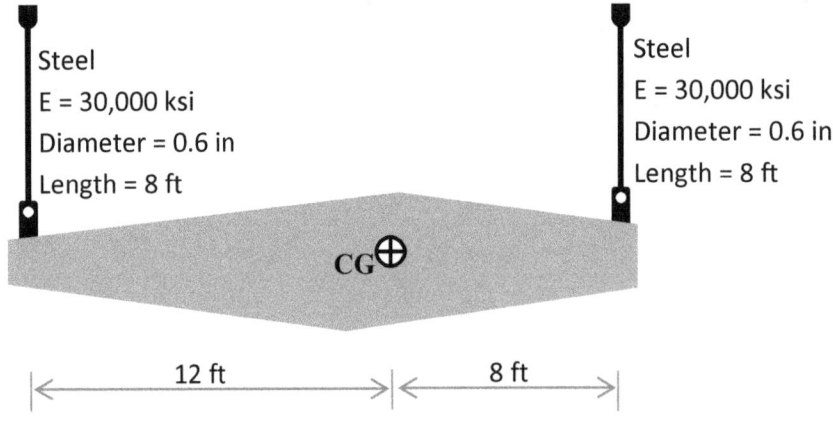

AM006

A single story building frame is subject to dead load (D), live load (L) and wind load (W) as indicated. The wind load is reversible (i.e. it can act from the left or from the right). The footings are being investigated for the possibility of uplift.

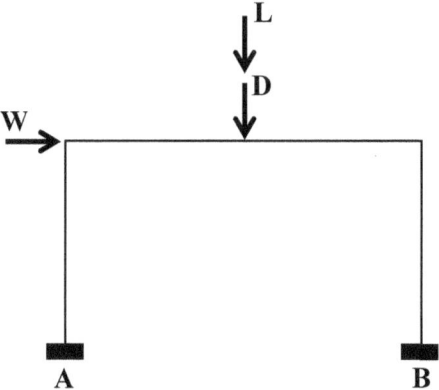

The load combination that should be used is:

A. D
B. D + L
C. D + L + W
D. D + W

AM007

A cantilever retaining wall is supported by a 3-foot thick footing as shown. The drains behind the wall become clogged and groundwater rises to the top of the horizontal backfill. The total horizontal earth pressure resultant (lb/ft) acting on the retaining wall is most nearly:

A. 3,110
B. 10,110
C. 13,220
D. 16,590

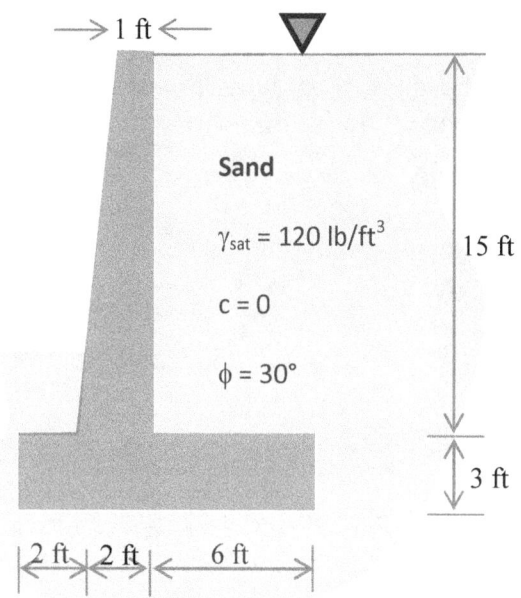

AM008

The particle size distribution curve for a soil sample is shown below. The curvature coefficient (Hazen) is most nearly:

    A. 1.25
    B. 1.90
    C. 2.50
    D. 4.55

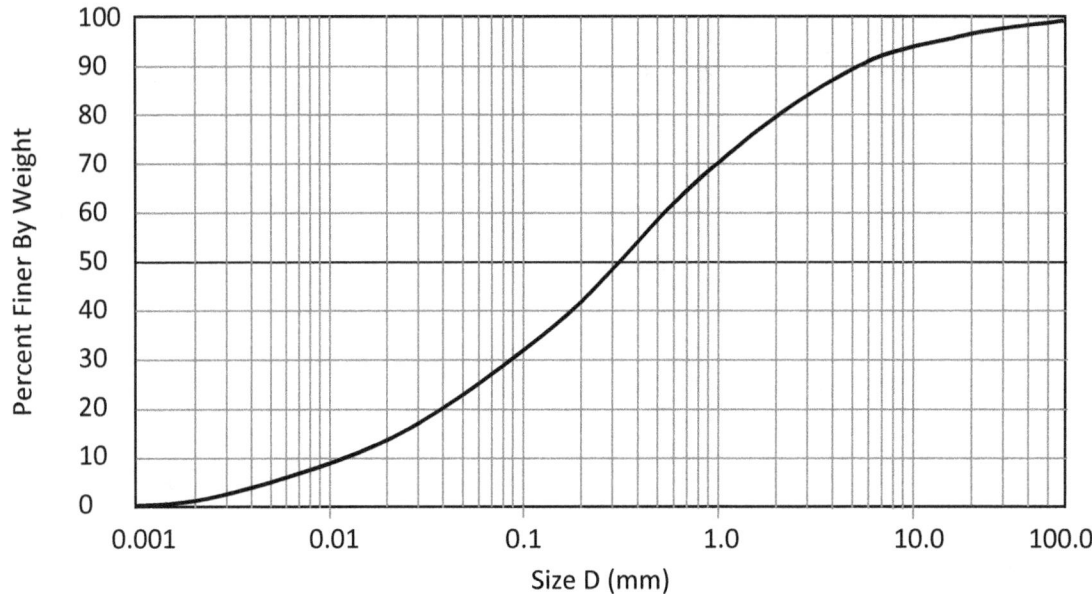

AM009

Which of the following soil types is most likely to respond to effective compaction under vibratory plate compaction?

    A. SM
    B. GC
    C. SP
    D. GW

AM010

A sample of soil has the following characteristics:

   Void ratio = 0.45

   Specific gravity of solids = 2.65

   Water content = 16%

The unit weight (lb/ft^3) of the soil is most nearly:

   A. 120

   B. 125

   C. 132

   D. 137

AM011

A plate load test is conducted on a sandy soil stratum. The plate is 12 in x 12 inch. A concentric static load of 6000 lbs is applied to the plate. The soil has the following properties:

   Unit weight = 125 lb/ft^3

   Moisture content = 15%

   Void Ratio = 0.50

   Modulus of elasticity = 300 kip/ft^2

   Coefficient of subgrade reaction = 500 lb/in^3

The vertical settlement (inches) due to the application of the load is most nearly:

   A. 0.05

   B. 0.08

   C. 0.10

   D. 0.15

AM012

Which of the following statements is true?

I. The effective stress is a moist soil immediately above the water table is calculated using the buoyant unit weight

II. The effective stress in a soil experiencing seepage flow can be either more or less than the effective stress with no seepage

III. The effective stress in a soil represents the pressure exerted by the pore water on the surface of the soil particles

IV. Increase in total stress in a soil is responsible for consolidation settlement

   A. II only

   B. II and IV only

   C. II, III and IV only

   D. I and III

AM 013

A circular footing (diameter = 36 inches) supports a column as shown. The minimum required factor of safety against ultimate bearing capacity = 3.0. The maximum permitted column load (kips) is most nearly:

    A. 47
    B. 52
    C. 57
    D. 62

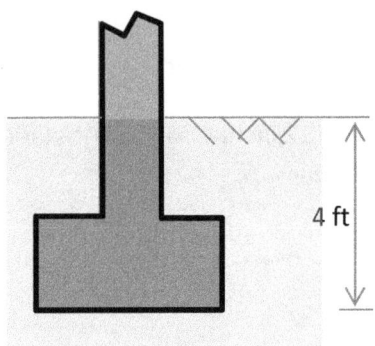

γ = 125 pcf
w = 12%
ϕ = 35°

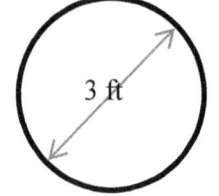

AM014

In the movement of material on a slope, creep is typically caused by

    A. rapid movement of soil downslope caused by gravity
    B. slow movement of solid rock downslope caused by gravity
    C. rapid movement of ice and snow downslope
    D. slow movement of soil downslope caused by contraction and expansion forces

AM015

The design target 28-day compressive strength of concrete for a bridge construction project is 6000 psi. Concrete samples will be tested as part of the QA/QC efforts that are part of the project. The sample standard deviation is 675 psi.

The minimum required average compressive strength (psi) to be used in concrete mix design is most nearly:
- A. 6500
- B. 6750
- C. 7000
- D. 7250

AM016

Which of the following is typically found on a boring log?
I. Water table elevation
II. Density and water content
III. Relative compaction
IV. USCS classification
V. CPT results

- A. I, III and V
- B. all of them
- C. I, IV and V
- D. I and V

AM 017

Which of the following is/are NOT an example of soil testing?
I. Ground Penetrating Radar
II. Nuclear density test
III. Brinell hardness test
IV. Liquid Penetrant Test
V. Cone penetrometer test

- A. II, III and V
- B. III and IV
- C. IV and V
- D. II and V

AM018

A charcoal filter is used for particulate removal from water by adsorption. The filter material has the following characteristics:

  Filter thickness = 12 inches
  Effective hydraulic conductivity = 2×10^{-3} ft/sec
  Porosity = 0.40
  Critical velocity (breakthrough) through the filter = 0.2 inch/sec.

If the desired factor of safety (against filter breakthrough) is 5.0, the maximum head difference (inches) that should be maintained across the filter is most nearly:

  A. 8
  B. 12
  C. 16
  D. 20

AM019

Normal depth of flow occurs in a rectangular open channel. The following data is given:

  Flow rate = 1200 cfs
  Manning's roughness coefficient = 0.015
  Bottom width of = 10 ft
  Longitudinal slope = 1%

The Froude number is most nearly:

  A. 0.5
  B. 1.4
  C. 3.9
  D. 7.2

AM020

A 24 inch diameter cast-iron pipe (friction factor = 0.022) conveys water at a flow rate = 25 ft³/sec. Water temperature is 60°F. Two points along the pipe (marked A and B) are separated by a distance of 1200 ft as shown. The average slope along the pipe is 1%. The pressure loss due to friction (lb/in²) is most nearly:

    A. 0.18
    B. 0.21
    C. 0.34
    D. 0.43

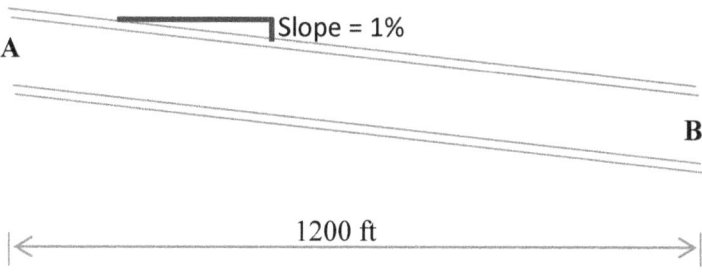

AM021

A watershed is subdivided into 5 distinct land-use/land-cover as summarized in the table below. The intensity-duration-frequency curves are synthesized from regional storm records.

Region	Land use/Land Cover	Area (acres)	Rational Coefficient	Time of concentration (min)
A	Wooded and forested	50	0.20	40
B	Residential lots	65	0.55	30
C	Parking lots	20	0.85	25
D	Lawns	240	0.30	35

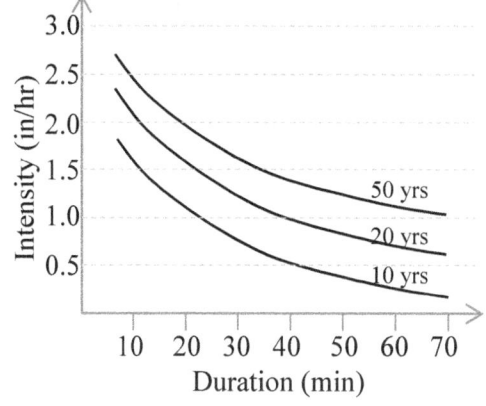

The discharge (ft³/sec) due to a 50-year storm is most nearly:

    A. 70
    B. 110
    C. 190
    D. 260

AM022
Which of the following statements is/are FALSE?
I. Flood stage (elevation) increases with an increase in the return period.
II. Annual probability of occurrence of a 20 year storm is exactly half of that for a 10 year storm
III. The 50-year flood is guaranteed to occur at least once in a 100 year interval
IV. Return period of design event does not directly impact the design

   A. I and III
   B. II and III only
   C. III only
   D. III and IV

AM023
The primary difference between retention ponds and detention ponds is
   A. Detention ponds are of larger depth than retention ponds
   B. Retention ponds always have water while detention ponds have water only periodically
   C. Settlement of coarse solids occurs in retention ponds while a detention pond achieves settlement of finer particles
   D. Detention ponds serve a hydraulic function, whereas retention ponds achieve a water quality management function

AM024
What is the frictional head loss (feet) is a 24 inch diameter concrete pipe over a length of 2400 ft carrying a flow rate of 10,000 gal/min. Assume a Hazen-Williams roughness coefficient of 110.
   A. 10
   B. 15
   C. 20
   D. 25

AM025

A stream with a tributary watershed of 180 acres has a gaging station where discharge is recorded. The table below shows a time vs. discharge record following a 2 hour rainfall event.

Time (hr)	0	1	2	3	4	5	6
Discharge (cfs)	42	70	138	208	88	66	40

The peak discharge of the 1-hour unit hydrograph (cfs per inch) is most nearly

    A. 85
    B. 75
    C. 65
    D. 45

AM026

A parabolic vertical curve of length = 1200 ft must connect a tangent of slope − 4% to +5% to another of slope +5% as shown below. The two tangents intersect at a point at station 11 + 45.20 and at elevation 310.56 ft. The elevation of the low point on the curve (feet) is most nearly

    A. 323.03
    B. 323.89
    C. 324.16
    D. 325.15

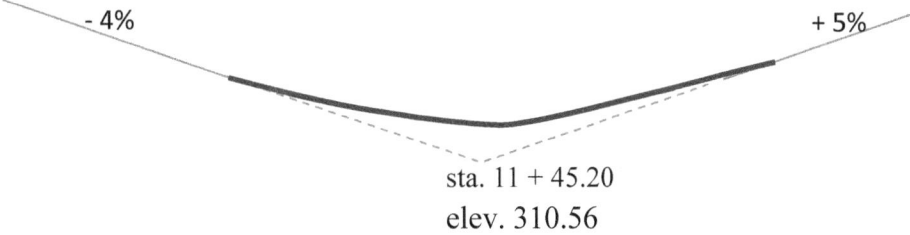

sta. 11 + 45.20
elev. 310.56

AM027

A circular horizontal curve connects a back tangent with bearing N62°E with a forward tangent with bearing S33°E as shown. The degree of curve = 5° The chord distance (feet) from PC the PT is most nearly:

    A. 1410
    B. 1550
    C. 1620
    D. 1740

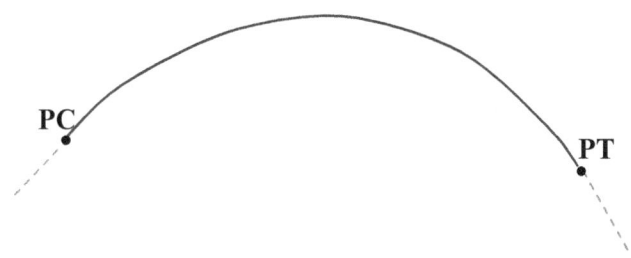

AM028

The average density of vehicles in a freeway lane is 30 vehicles per mile. If the average speed is 45 mph, the average headway (seconds) is most nearly:

    A. 2.7
    B. 2.4
    C. 2.1
    D. 1.8

AM029

The table below shows cross section areas of cut and fill recorded at 5 stations spaced at 100 ft. Shrinkage is 15%

Station	Area (ft²) CUT	Area (ft²) FILL
0 + 0.00	245.0	423.5
1 + 0.00	312.5	176.3
2 + 0.00	111.5	303.0
3 + 0.00	234.5	188.4
4 + 0.00	546.2	514.5

If the ordinate of the mass diagram at station 0 + 00 is +400 yd³, the ordinate (yd³) at station 4 + 00 is most nearly

    A. – 650
    B. – 1,450
    C. – 27,920
    D. +28,720

AM030

While conducting a site survey, a benchmark elevation of 154.45 ft above sea level is established. A level and rod arrangement measures the following:

    Rod reading at benchmark = 7.85 ft
    Rod reading at station A = 8.92 ft

The elevation of station A (feet) is most nearly:

    A. 171.22
    B. 155.52
    C. 153.38
    D. 137.68

AM031

A crane with a 40 ft boom is used to lift a load (W = 12 tons) as shown. The total weight of the crane and ballast is 4.5 tons acting at the effective location indicated as CG on the figure. The weight of the boom is 800 lb. Each outrigger leg is supported by a circular pad with a diameter = 3 feet.

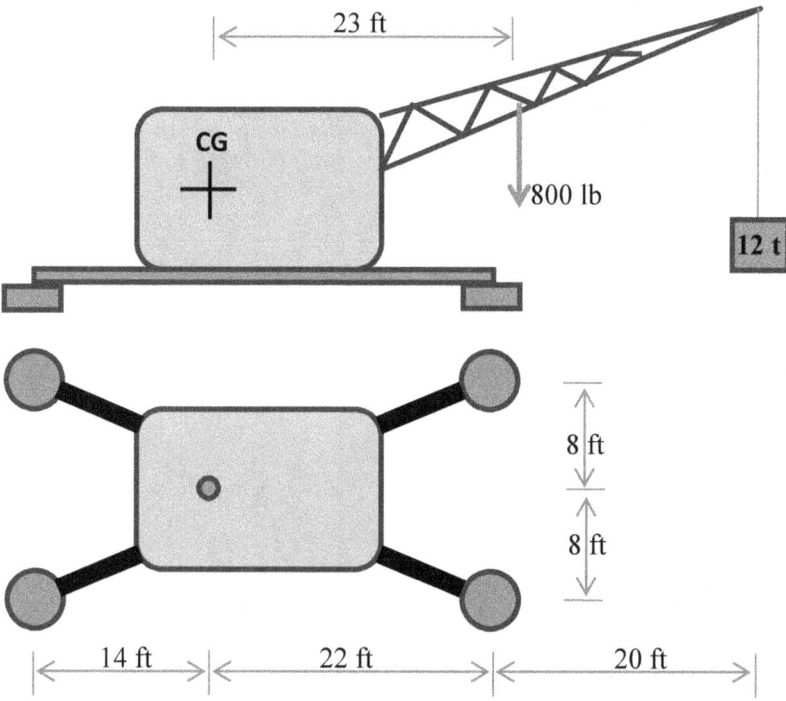

If the ultimate bearing capacity of the surface soil is 9000 psf, the factor of safety against bearing capacity failure is most nearly:

    A. 1.5
    B. 2.0
    C. 2.5
    D. 3.0

AM032

A freestanding masonry wall of height 16 feet is being constructed. During construction, the width of the 'restricted zone' should extend:

    A. 18 feet on one side of the wall
    B. 18 feet on both sides of the wall
    C. 20 feet on one side of the wall
    D. 20 feet on both sides of the wall

AM033

A bridge consists of a concrete deck supported by 4 equally spaced girders as shown below. Repair of piers requires that a temporary jacking tower be located adjacent to each pier and a line of 6 hydraulic jacks be used to raise the girders using a jacking beam. The total dead load incident to the jacking beam during the jacking operations is 243 kips.

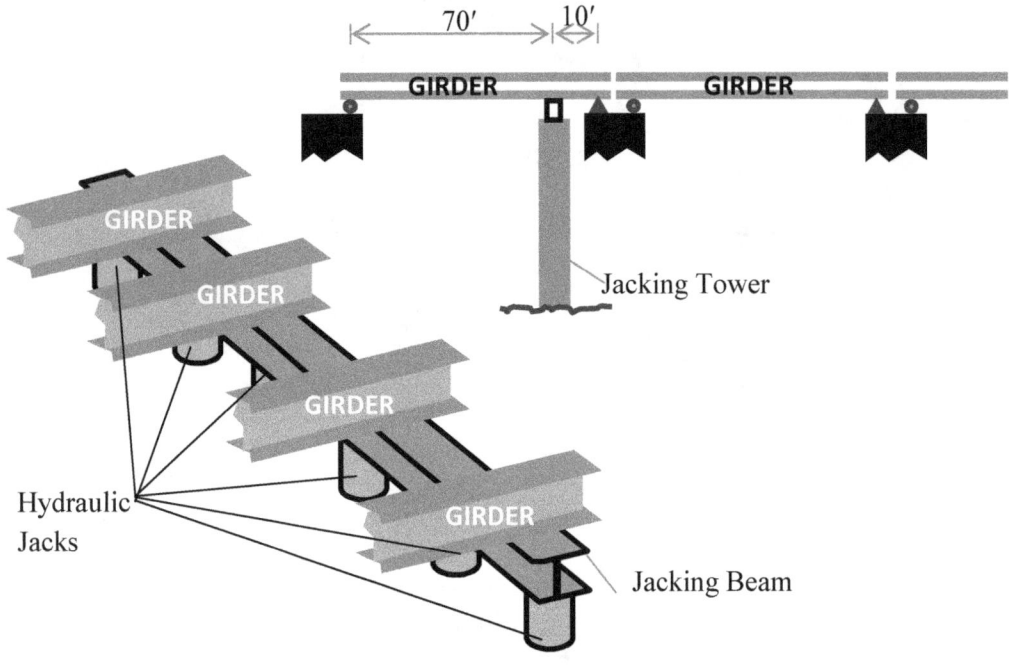

Using a factor of safety of 1.8, the minimum required capacity (tons) of the hydraulic jacks is most nearly:

    A. 40
    B. 32
    C. 24
    D. 18

AM034

A two-story building has a rectangular footprint 120 ft x 85 ft (outer dimensions). The overall building height is 25 ft. The outer walls are to be double-wythe brick walls made with standard bricks (3.625 in x 2.25 in x 8 in) with 3/8 inch mortar thickness. A vertical section through a section of the wall is shown below. The number of bricks needed to build the double wythe wall is most nearly

    A. 125,000
    B. 135,000
    C. 155,000
    D. 168,000

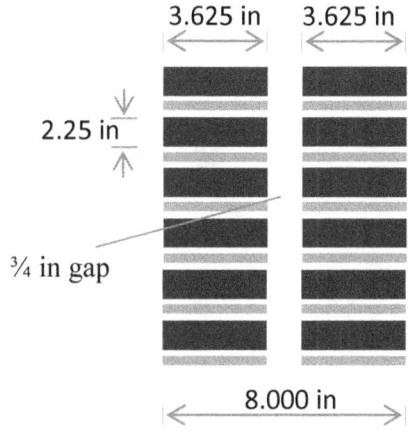
VERTICAL SECTION THROUGH WALL

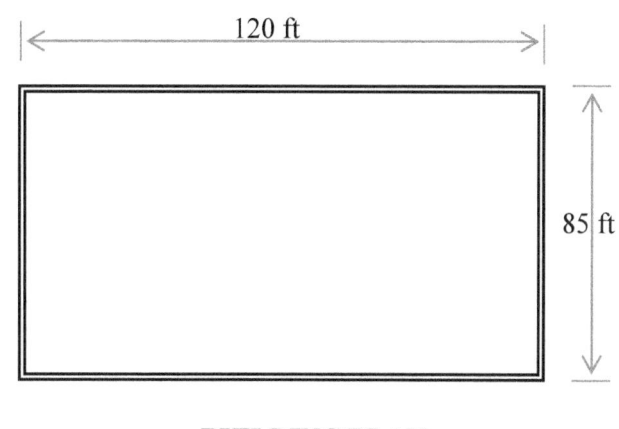
BUILDING PLAN

AM035

A 4-member crew pours concrete at the rate of 20 yd^3/hr. Crew labor rate is $105/hour. Equipment rental costs $54/8-hour day. Assuming an 8-hour workday, what is the total cost of labor + equipment for a job that requires pouring of 755 yd^3 of concrete?

    A. $4,300
    B. $5,600
    C. $6,910
    D. $7,450

AM036

The activity on node network for a project is as shown below. All relationships are finish to start unless otherwise indicated. Numbers in parentheses are activity durations in weeks. The minimum time to complete the project (weeks) is:

    A. 21
    B. 26
    C. 29
    D. 31

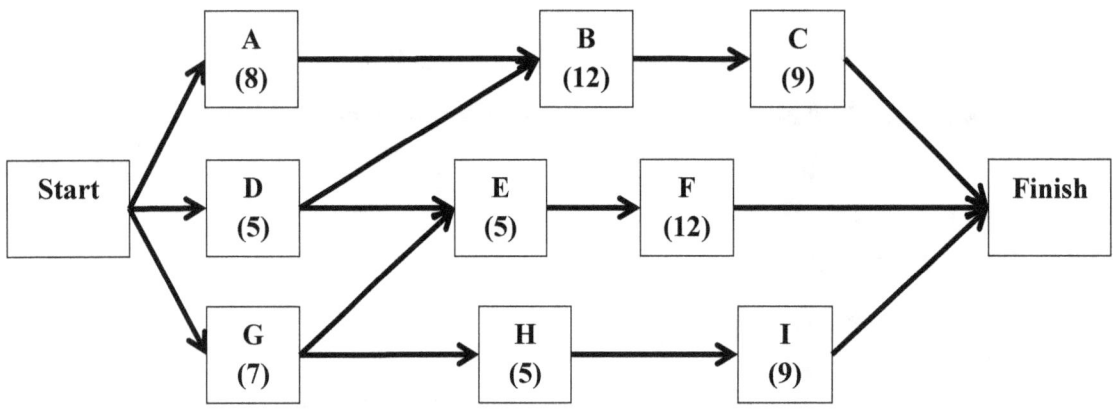

AM037

During exercising project controls at a certain time for a project, the project manager computes the following:

BCWS = $435,000
ACWP = $510,000
BCWP = $488,000

Which of the following is true?

    A. The project is behind schedule and over budget
    B. The project is ahead of schedule and under budget
    C. The project is behind schedule but under budget
    D. The project is ahead of schedule but over budget

AM038

During repaving of a highway segment, project specifications dictate that a buffer zone of width 12 feet on either side of the 65 foot wide roadway be subject to clearing and grubbing. Project limits (measured along baseline curve along centerline of roadway) are from station 5 + 05.25 to 25 + 31.20. Unit cost for clearing and grubbing is $0.20 per SF.

The total line item cost for "Clearing and Grubbing" is most nearly:

    A. $10,000
    B. $12,000
    C. $14,000
    D. $16,000

AM039

Which of the following are considered acceptable measures for ensuring trench safety?

I. Keeping the soil dry
II. Providing benches
III. Shielding

    A. I and III
    B. II and III
    C. I and III
    D. I, II and III

AM040

A longitudinal section of a pipe segment between two manholes MH-1 and MH-2 is shown below. The longitudinal slope of the pipe segment is most nearly:

　　A. 1.2%
　　B. 1.5%
　　C. 1.8%
　　D. 2.1%

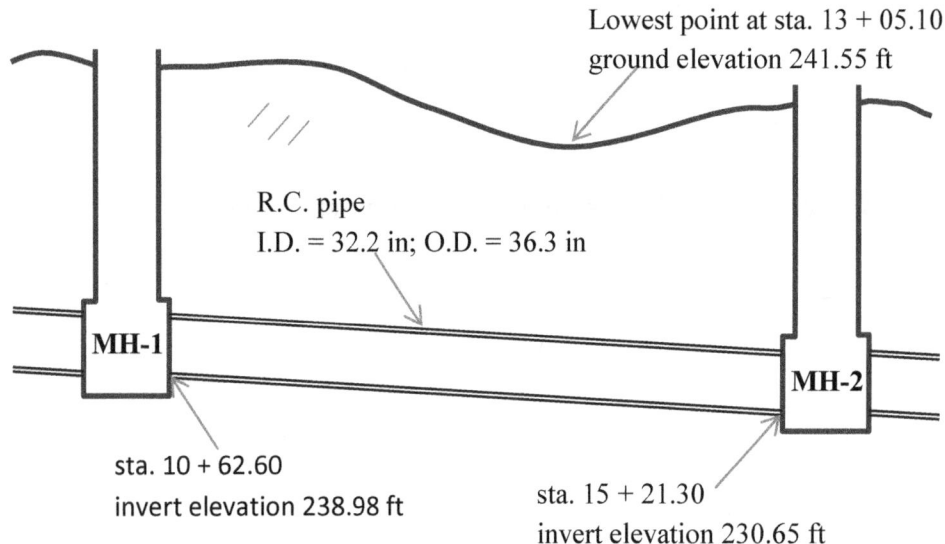

THIS IS THE END OF THE BREADTH EXAM

ANSWER SHEET – BREADTH (AM) EXAM

AM001	AM009	AM017	AM025	AM033
AM002	AM010	AM018	AM026	AM034
AM003	AM011	AM019	AM027	AM035
AM004	AM012	AM020	AM028	AM036
AM005	AM013	AM021	AM029	AM037
AM006	AM014	AM022	AM030	AM038
AM007	AM015	AM023	AM031	AM039
AM008	AM016	AM024	AM032	AM040

This page intentionally left blank

SOLUTIONS TO BREADTH EXAM
FOR THE
CIVIL PE EXAM

ANSWER KEY – BREADTH (AM) EXAM

AM001	D	AM009	D	AM017	B	AM025	A	AM033	A
AM002	B	AM010	C	AM018	A	AM026	B	AM034	B
AM003	B	AM011	B	AM019	B	AM027	B	AM035	A
AM004	C	AM012	A	AM020	D	AM028	A	AM036	C
AM005	A	AM013	C	AM021	C	AM029	A	AM037	D
AM006	D	AM014	D	AM022	D	AM030	C	AM038	A
AM007	C	AM015	C	AM023	B	AM031	D	AM039	B
AM008	A	AM016	D	AM024	C	AM032	D	AM040	C

Solution AM001

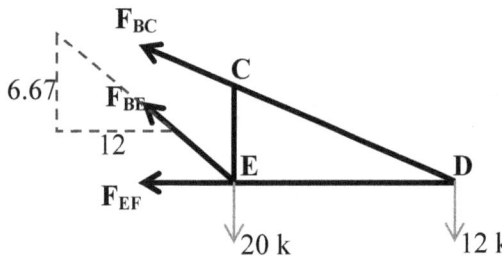

Making a section through BC, BE and EF and then taking moments about D (members BC and EF pass through point D, and they will be absent from this equation)

$$\sum M_D = \frac{6.67}{13.73} F_{BE} \times 12 - 20 \times 12 = 0 \Rightarrow F_{BE} = +41.17 \, k$$

Answer is D

Solution AM002

At the support, the internal forces are:

Shear Force: V = 10 sin 40 = 6.428 kips

Axial Force: A = 10 cos 40 = 7.660 kips

Bending Moment: M = 10sin(40)x60 + 10cos(40)x20 = 538.9 kip-inch (clockwise)

Since the axial force is tensile, the maximum stress (tension) at the top edge is:

$$\sigma = \frac{P}{A} + \frac{My}{I} = \frac{10}{6 \times 8} + \frac{538.9 \times 4}{\frac{1}{12} \times 6 \times 8^3} = 0.21 + 8.42 = 8.63$$

Answer is B

Solution AM003

There are 5 loading zones on the beam – the load function on them, left to right, are w = 0, w = 0, w = 0, w = 0 and w = constant respectively. As a result, the bending moment function is M = linear, linear, linear, linear and quadratic, linear respectively. Also, because of the downward loading on each span, there is expected to be negative bending moment (tension on top) over and adjacent to the supports. This eliminates choices A and C. There is a change in shear force at the center of span 2, which means the bending moment diagram must show a change of slope at this point. This eliminates D

Answer is B

Solution AM004

For a cantilever beam under a point load at the free end, the maximum deflection is given by:
$$\Delta_{max} = \frac{PL^3}{3EI} = \frac{25 \times 180^3}{3 \times 29000 \times 890} = 1.883 \ in$$

Answer is C

Solution AM005

Since this is a determinate structure, cable tensions can be calculated easily from equilibrium equations. Taking moments about the cable on the left (T_1), we can calculate the tension in the cable on the right (T_2)

$\sum M_{T_1} = 0 \Rightarrow 20000 \times 12 - T_2 \times 20 = 0 \Rightarrow T_2 = 12000 \ lb; T_1 = 8000 \ lb$

Elongation of cable 1: $\Delta_1 = \frac{TL}{AE} = \frac{8 \times 96}{\frac{\pi}{4} \times 0.6^2 \times 30,000} = 0.091 \ in$

Elongation of cable 2: $\Delta_2 = \frac{TL}{AE} = \frac{12 \times 96}{\frac{\pi}{4} \times 0.6^2 \times 30,000} = 0.136 \ in$

However, the deflection of the center of gravity will be (linearly averaged between points 1 and 2)
$$\Delta_{CG} = \frac{1}{3}\Delta_1 + \frac{2}{3}\Delta_2 = 0.121 \ in$$

Answer is A

Solution AM006

If the design objective is to investigate the possibility of uplift at a support (nearside support A is most vulnerable), then the chosen load combination should include the wind load (which creates the possibility of uplift) but exclude the live load (L), because it has the potential to counter the uplift caused by the wind load. However, we have no such discretion about the dead load, which must appear at full strength.

Answer is D

Solution AM007
For ϕ = 30, active earth pressure coefficient: $K_a = \frac{1-\sin\phi}{1+\sin\phi} = 0.333$

At the base of the footing (depth below ground surface = 18 ft), the effective earth pressure: $K_a \gamma_{sub} H = 0.333 \times (120 - 62.4) \times 18 = 345.3 \ psf$
At the base of the footing (depth below water table = 18 ft), the hydrostatic pressure: $\gamma_w H = 62.4 \times 18 = 1123.2 \ psf$
Total pressure at bottom of footing = 1468.5 psf

Total active resultant = 0.5x1468.5x18 = 13,217 lb/ft

Answer is C

Solution AM008
From the particle size distribution curve: D_{10} = 0.012 mm, D_{30} = 0.09 mm, D_{60} = 0.55 mm
Hazen coefficient of curvature: $C_c = \frac{D_{30}^2}{D_{10} D_{60}} = \frac{0.09^2}{0.012 \times 0.55} = 1.23$

Answer is A

Solution AM009
Coarser grained soils with a cascade of sizes (wide band of particle size distribution) and fewer fines tend to compact well under vibration. This would indicate a well-graded sand.

Answer is D

Solution AM010
The degree of saturation is given by: $S = \frac{wG_s}{e} = \frac{0.16 \times 2.65}{0.45} = 0.942$

The total (moist) unit weight is given by: $\gamma = \frac{G_s + Se}{1+e} \gamma_w = \frac{2.65 + 0.942 \times 0.45}{1 + 0.45} \times 62.4 = 132.3 \ pcf$

Answer is C

The following approach could also have been taken (if convenient formulas were not available)
Consider Volume of solids = 1.0 ft³
Therefore, volume of voids = 0.45 ft³

Total volume = 1.45 ft³
Weight of solids = 2.65 x 62.4 x 1.0 = 165.36 lb
Weight of water = 0.16 x 165.36 = 26.46 lb
Total weight = 191.82 lb
Total unit weight = 191.82 ÷ 1.45 = 132.3 pcf

Solution AM011

Bearing pressure applied to the soil: $q = \frac{P}{B^2} = \frac{6000}{12^2} = 41.67 \; psi$

Settlement: $\Delta = \frac{q}{k} = \frac{41.67}{500} = 0.083 \; in$

Answer is B

Solution AM012

I is false because above the water table, the total unit weight and not the buoyant unit weight should be used.
III is false because the effective stress is the pressure exerted by the soil particles on each other.
IV is false because increase in effective stress (rather than total stress) is responsible for consolidation settlement.

Answer is A

Solution AM013

The Terzaghi bearing capacity factors for ϕ = 35° are N_c = 58.1, N_q = 41.8 and N_γ = 46.2. Based on that, and D = 4 ft and B = 3 ft, the ultimate bearing capacity for a circular footing, using Terzaghi's theory, is given by:

$$q_{ult} = 1.3cN_c + \gamma D N_q + 0.3\gamma B N_\gamma = 0 + 125 \times 4 \times 41.8 + 0.3 \times 125 \times 3 \times 46.2 = 26{,}098 \; psf$$

Therefore, allowable bearing capacity = 8,699 psf.

$$\frac{P}{A} + \gamma D \leq 8{,}699 \Rightarrow \frac{P}{A} \leq 8{,}699 - 125 \times 4 = 8{,}199 \Rightarrow P \leq 8{,}199 \times \frac{\pi \times 3^2}{4} = 57{,}955 \; lb$$

Answer is C

Solution AM014

Creep can be caused by the expansion of materials such as clay when they are exposed to water. Clay expands when wet, pushing the material downhill. When the soil dries, contraction results in consolidation of the soil in the new position.

Answer is D

Solution AM015

For $f_c' > 5000$ psi, the required compressive strength f_{cr}' must be the greater of

$f_{cr}' = f_c' + 1.34s_s = 6000 + 1.34 \times 675 = 6,905$ psi and
$f_{cr}' = 0.9f_c' + 2.33s_s = 0.9 \times 6000 + 2.33 \times 675 = 6,973\ psi$ psi
Therefore, f_{cr}' must be greater than 6973 psi

Answer is C

Solution AM016

A boring log shows water table, in situ results such as CPT and SPT, Atterberg limits.

Answer is D

Solution AM017

GPR, nuclear density test, and the cone penetrometer test are all soil tests. Liquid penetrant test is typically used to detect surface defects of plastics, metals and ceramics. The Brinell hardness test is used to measure surface hardness of metals. Therefore III and IV are not used for soils.

Answer is B

Solution AM018

Breakthrough velocity = 0.2 in/sec
Hydraulc conductivity = 0.002 ft/sec = 0.024 in/sec
Maximum velocity of flow = 0.2 ÷ 5 = 0.04 in/sec

Seepage velocity: $v = \frac{Ki}{n} = \frac{K\left(\frac{H}{L}\right)}{n} = \frac{KH}{nL} \leq 0.04 \Rightarrow H \leq \frac{0.04 \times 0.4 \times 12}{0.024} = 8\ in$

Answer is A

Solution AM019

Flow rate parameter: $K = \frac{Qn}{1.486b^{8/3}S^{1/2}} = \frac{1200 \times 0.015}{1.486 \times 10^{8/3} 0.01^{1/2}} = 0.2610$

For this K and for m = 0 (rectangular channel), d/b = 0.616
Depth of flow = 0.616x10 = 6.16 ft (for a rectangular channel, this is also hydraulic depth)
Velocity: $V = \frac{Q}{A} = \frac{1200}{10 \times 6.16} = 19.48\ fps$

Froude number: $Fr = \frac{V}{\sqrt{gd_h}} = \frac{19.48}{\sqrt{32.2 \times 6.16}} = 1.38$

Answer is B

Solution AM020

By the continuity principle, the velocity must remain constant in the pipe (diameter unchanged).

Velocity: $V = \frac{Q}{A} = \frac{25}{\frac{\pi}{4} \times 2^2} = 7.96 \, fps$

According to Bernoulli's Principle:

$$\frac{p_A}{\gamma} + \frac{V_A^2}{2g} + z_A - h_f = \frac{p_B}{\gamma} + \frac{V_B^2}{2g} + z_B \Rightarrow \frac{p_A - p_B}{\gamma} = h_f + z_B - z_A + \frac{V_B^2 - V_A^2}{2g} = h_f - 12$$

Elevation difference: $z_B - z_A$ = -0.01x1200 = -12 ft

Head loss (using Darcy-Weisbach theory): $h_f = f \frac{L}{D} \frac{V^2}{2g} = 0.022 \times \frac{1200}{2} \times \frac{7.96^2}{2 \times 32.2} = 12.99 \, ft$

Therefore, pressure loss = 12.99 – 12 = 0.99 ft, which is equivalent to 0.99x62.4 = 61.6 psf = 0.43 psi

Answer is D

Solution AM021

The composite Rational coefficient is: $\bar{C} = \frac{\sum C_i A_i}{\sum A_i} = \frac{0.2 \times 50 + 0.55 \times 65 + 0.85 \times 20 + 0.3 \times 240}{375} = 0.36$

Governing time of concentration = 40 minutes. From the I-D-F curves, for a return period = 50 years, the design intensity = 1.4 in/hr

Discharge; $Q = CiA = 0.36 \times 1.4 \times 375 = 189 \, ac - \frac{in}{hr} = 190.5 \, ft^3/sec$

Answer is C

Solution AM022

As the return period increases, the annual probability of occurrence decreases. This corresponds to a stronger event (higher flood elevation). I is correct

Annual probability of 20 year storm = 1/20 = 0.05, while annual probability of 10 year storm = 1/10 = 0.10. So, II is correct

III is incorrect because even though it is small, there still exists a non-zero probability that a 50 year flood will NOT occur in the next 100 years

IV is incorrect because the magnitude of the design event is directly influenced by the return period

Answer is D

Solution AM023

A. is not necessarily true. C is exactly reverse – because of a continuous presence of water in retention ponds, finer solids, which take longer to settle, are removed to a greater extent than in detention ponds.

D is not true – even though the degree may vary, both detention and retention ponds serve both a hydraulic (flood control) as well as a water quality (solids removal) function.

Answer is B

Solution AM024

Head loss (feet) is given by: $h_f = \dfrac{5.862 \times 10^{-5} \times Q_{gpm}^{1.85} \times L_{ft}}{C^{1.85} \times D_{ft}^{4.865}} = \dfrac{5.862 \times 10^{-5} \times 10000^{1.85} \times 2400}{110^{1.85} \times 24^{4.865}} = 20.3\ ft$

Answer is C

Solution AM025

First, the base flow needs to be separated from the stream discharge. This will be accomplished by subtracting 42 cfs from the recorded discharge values. The last value is just made zero instead of -2

Time (hr)	0	1	2	3	4	5	6	
Discharge (cfs)	0	28	96	166	46	24	0	Sum = 360 cfs

The area under the Q-t curve is estimated using a histogram.

$$V = \sum (Q \Delta t) = \Delta t \sum Q = 3600\ sec \times 360\ cfs = 1{,}296{,}000\ ft^3$$

The average depth of runoff: $d = \dfrac{V}{A} = \dfrac{1{,}296{,}000}{180 \times 43560} = 0.165\ ft = 1.98\ in$

Therefore, the peak discharge of the UNIT hydrograph is 166 ÷ 1.98 = 83.8 cfs/in

Answer is A

Solution AM026

Rate of gradient change: $R = \dfrac{G_2 - G_1}{L} = \dfrac{+5 - (-4)}{12} = 0.75\ \%/sta$

Elevation of PVC: $y_{PVC} = y_{PVI} - G_1 \dfrac{L}{2} = 310.56 - (-4) \times 6 = 334.56\ ft$

Location of low point is: $x_{LP} = -\dfrac{G_1}{R} = -\dfrac{-4}{0.75} = +5.33\ sta$

At this location, elevation of the curve: $y = y_{PVC} + G_1 x + \dfrac{1}{2} R x^2 = 334.56 + (-4) \times 5.33 + 0.5 \times 0.75 \times 5.33^2 = 323.89\ ft$

Answer is B

Solution AM027

Azimuth of forward tangent = 180 − 33 = 147°
Azimuth of back tangent = 62°
Deflection angle I = 147 − 62 = 85°
Radius: $R = \frac{5729.578}{D} = \frac{5729.578}{5} = 1145.92 \, ft$
Long chord length: $2R \sin \frac{I}{2} = 2 \times 1145.92 \times \sin 42.5 = 1548.34 \, ft$

Answer is B

Solution AM028

Density D = 30 veh/mile
Speed S = 45 miles/hour
Flow rate = SD = 1350 veh/hour
Headway = 3600 seconds/hour ÷ 1350 vehicles/hour = 2.67 sec/vehicle

Answer is A

Solution AM029

Using the average end-area method, we calculate the total volume of cut between stations 0+ 0.00 and 4 + 0.00 as:
$V_C = \frac{100}{2}[245.0 + 546.2 + 2 \times (312.5 + 111.5 + 234.5)] = 105{,}410 \, ft^3$
Similarly, the unadjusted volume of fill between stations 0+0 and 4+0 is:
$V_F = \frac{100}{2}[423.5 + 514.5 + 2 \times (176.3 + 303.0 + 188.4)] = -113{,}670 \, ft^3$
Adjusting the fill volume for shrinkage (shrinkage factor = 0.85), we have
V_F = -113,670 ÷ 0.85 = - 133,729.4 ft³

Net earthwork between 0+0.00 and 4+0.00 = 105,410 − 133,729.4 = - 28,319.4 ft³ = - 1048.87 yd³
Mass diagram ordinate at station 4 + 0.00 is +400 − 1048.87 = - 648.87

Answer is A

Solution AM030

Rod reading at benchmark = 7.85. Therefore, height of instrument = 154.45 + 7.85 = 162.30 ft

Rod reading at A = 8.92 ft. Therefore, elevation at A = height of instrument − rod reading = 162.30 − 8.92 = 153.38 ft

Answer is C

Solution AM031

Representing the system by a 2D free body diagram as below:

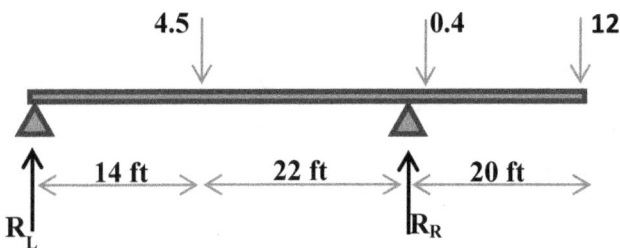

The largest soil reaction will be at the supports on the right. Taking moments about the left support:

$$\sum M_{R_L} = 4.5 \times 14 - 36 R_R + 0.4 \times 37 + 12 \times 56 = 0 \Rightarrow R_R = 20.83 \; tons$$

Therefore, reaction at each outrigger leg on the right = 10.42 tons = 20,827 lbs
Are of each pad (diameter = 3 ft) = 7.07 ft²
Bearing pressure = 2,946 psf

Factor of safety = 9,000/2,946 = 3.05

Answer is D

Solution AM032

According to the guidelines of the Standard Practice for Bracing Masonry Walls During Construction, the restricted zone should extend a width = H + 4 ft on either side of the wall.

Answer is D

Solution AM033

The total force on 6 jacks = 243 k
Assuming this load is equally shared by the 6 jacks, each jack needs to exert a force of 40.5 k = 20.3 tons

If a minimum FS = 1.8 is desired, the jacks must be rated at 20.3x1.8 = 36.5 tons

Answer is A

Solution AM034

The wall thickness is 8 inches. Therefore, centerline dimensions of the building wall are 119 ft 8 in x 84 ft 8 in. The perimeter of the building, along this centerline, is 408.67 ft. Since the building height is 25 ft, the total surface area (both wythes) = 2 x 408.67 x 25 = 20,434 ft²

Including a half-thickness of mortar (3/16 in) on all sides of brick faces, the equivalent wall surface area covered by each brick is (2.25+0.375) inch by (8.00+0.375) inch = 21.98 in² = 0.15267 ft²

Total number of bricks needed = 20,434 ÷ 0.15267 = 133,845

Answer is B

Solution AM035
Total time needed for the concrete pour = 755 ÷ 20 = 37.75 hours (use 38 hrs). For equipment rental, use 5 full days.

Labor cost = $105 x 38 = $3,990
Equipment rental cost = $54 x 5 = $270
Total cost = $4,260
Answer is A

Solution AM036
Start → A → B → C → Finish   29 weeks
Start → D → B → C → Finish   26 weeks
Start → D → E → F → Finish   22 weeks
Start → G → H → I → Finish   21 weeks
Start → G → E → F → Finish   24 weeks

Answer is C

Solution AM037
The cost performance index is calculated as
$$CPI = \frac{BCWP}{ACWP} = \frac{488,000}{510,000} < 1$$
Therefore, the project is currently over budget

The schedule performance index is calculated as
$$SPI = \frac{BCWP}{BCWS} = \frac{488,000}{435,000} > 1$$
Therefore, the project is currently ahead of schedule

Answer is D

Solution AM038
Length of zone = 2531.20 – 505.25 = 2025.95 ft
Width of zone = 24 ft
Area subject to clearing and grubbing = 48,622.8 ft^2
Cost = $9,724.56

Answer is A

Solution AM039

Sloping and benching and using shields are all acceptable methods for ensuring trench safety. Controlling soil moisture is not.

Answer is B

Solution AM040

Elevation difference between inverts = 238.98 – 230.65 = 8.33 ft
Horizontal distance = 1521.30 – 1062.60 = 458.70 ft
Slope = 8.33 ÷ 458.70 = 1.81%

Answer is C